SEA STARS
Pacific Coast Predators

RACHAEL L. THOMAS

Consulting Editor, Diane Craig, M.A./Reading Specialist

Super Sandcastle

An Imprint of Abdo Publishing
abdobooks.com

ABDOBOOKS.COM

Published by Abdo Publishing, a division of ABDO, PO Box 398166, Minneapolis, Minnesota 55439.
Copyright © 2020 by Abdo Consulting Group, Inc. International copyrights reserved in all countries.
No part of this book may be reproduced in any form without written permission from the publisher.
Super SandCastle™ is a trademark and logo of Abdo Publishing.

Printed in the United States of America, North Mankato, Minnesota
102019
012020

Design: Kelly Doudna, Mighty Media, Inc.
Production: Mighty Media, Inc.
Editor: Liz Salzmann
Cover Photographs: iStockphoto, Shutterstock Images
Interior Photographs: Getty Images/iStockphoto, pp. 8, 9, 10, 11, 14, 15, 19; Shutterstock Images,
pp. 3, 4, 5, 6, 7, 8, 9, 10, 11, 12, 13, 14, 15, 16, 17, 18, 19, 21, 22, 23; Wikimedia Commons-Oregon
State University, pp. 20, 21

Publisher's Cataloging-in-Publication Data
Names: Thomas, Rachael L., author.
Title: Sea stars: pacific coast predators/ by Rachael L. Thomas
Other title: pacific coast predators
Description: Minneapolis, Minnesota : Abdo Publishing, 2020 | Series: Animal eco influencers
Identifiers: ISBN 9781532191909 (lib. bdg.) | ISBN 9781532178634 (ebook)
Subjects: LCSH: Starfishes--Juvenile literature. | Sea stars--Juvenile literature. | Echinoderms--Juvenile
    literature. | Ocean animals--Juvenile literature. | Predatory animals--Behavior--Juvenile literature. |
    Animal ecology--Juvenile literature. | Wildlife habitats--Juvenile literature. |
Classification: DDC 593.93--dc23

Super SandCastle™ books are created by a team of professional educators, reading specialists, and
content developers around five essential components—phonemic awareness, phonics, vocabulary, text
comprehension, and fluency—to assist young readers as they develop reading skills and strategies and
increase their general knowledge. All books are written, reviewed, and leveled for guided reading, early
reading intervention, and Accelerated Reader™ programs for use in shared, guided, and independent
reading and writing activities to support a balanced approach to literacy instruction.

CONTENTS

ECO INFLUENCERS

What is an eco influencer? It is an animal that can change its ecosystem. All members of an ecosystem affect one another.

Parrotfish eat **algae** that grows on coral **reefs**. This keeps the reefs healthy.

Sea otters help shape their coastal ecosystem. They maintain **kelp** forests by eatlng sea urchins.

Ochre sea stars are eco **influencers**. They prey on mussels in tide pools. This makes room for other wildlife. Ochre sea stars are Pacific coast predators!

COASTAL RESIDENTS

There are about 2,000 kinds of sea stars. They live in oceans around the world. Ochre sea stars are one kind of sea star. They are found on the Pacific coast of **North America**.

At low tide, ochre sea stars are commonly seen in tide pools.

TIDE POOLS

The ochre sea star's **habitats** are affected by the ocean's tides. When the tide is high, the habitats are underwater. When the tide is low, tide pools form near the shore. These habitats are exposed to the sun and the air.

SEA STARS BY THE SEASHORE

Tide pools are busy saltwater **ecosystems**. They provide food and shelter for sea stars and other wildlife.

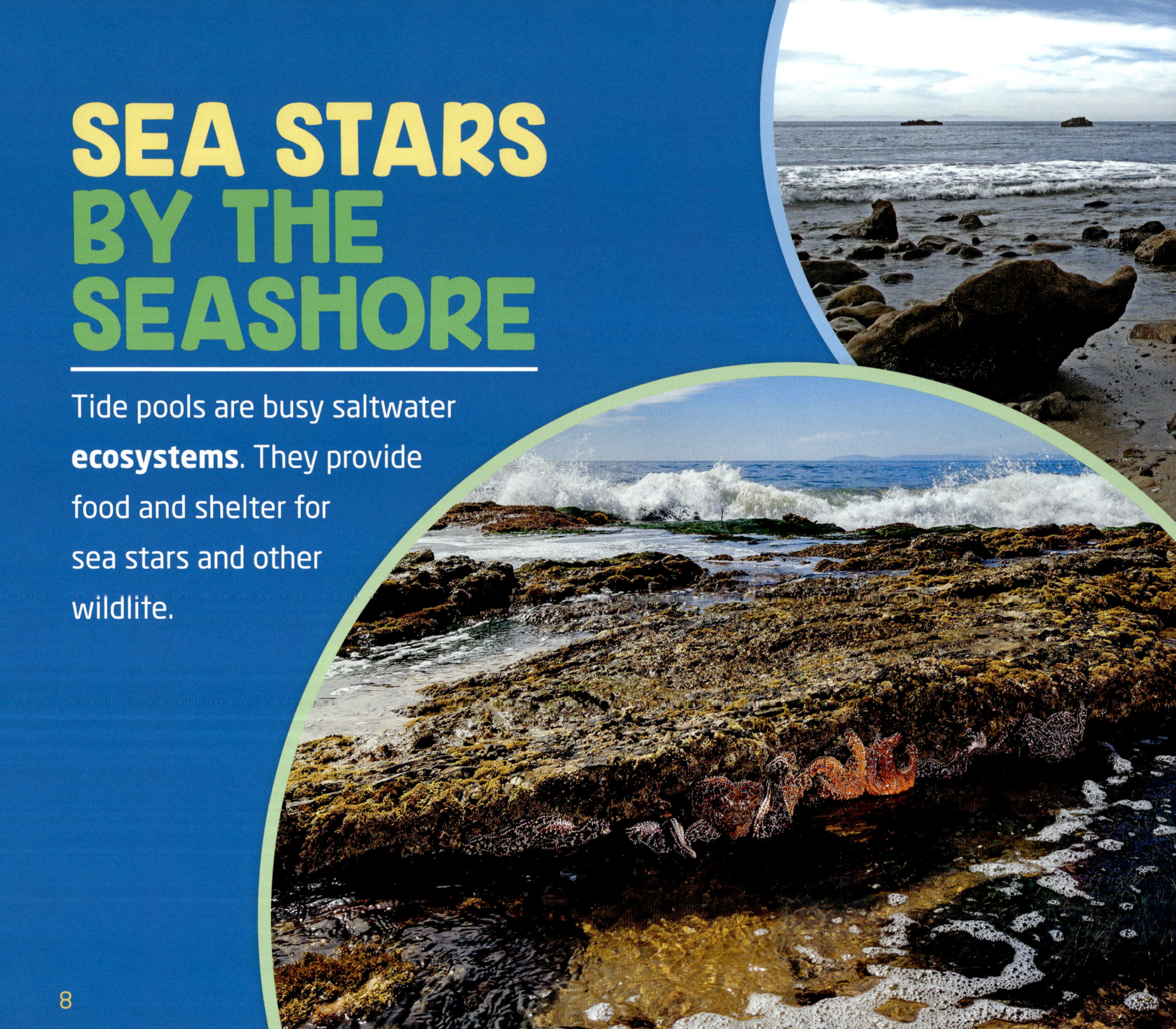

The ocean's waves bring **nutrients** and oxygen to tide pools.

Rocks provide shelter, surfaces to cling to, and cracks to hide in.

Shallow water lets sunlight reach the seabed. This helps **algae** and other plants grow. Healthy plant life means more food for sea animals.

THINK!
What do you like best about your home?

POKY PREDATORS

Ochre sea stars move very slowly. They may not seem very **dangerous**. But sea stars are serious predators! Barnacles, mussels, and clams live in tide pools. Ochre sea stars are built to prey on these animals.

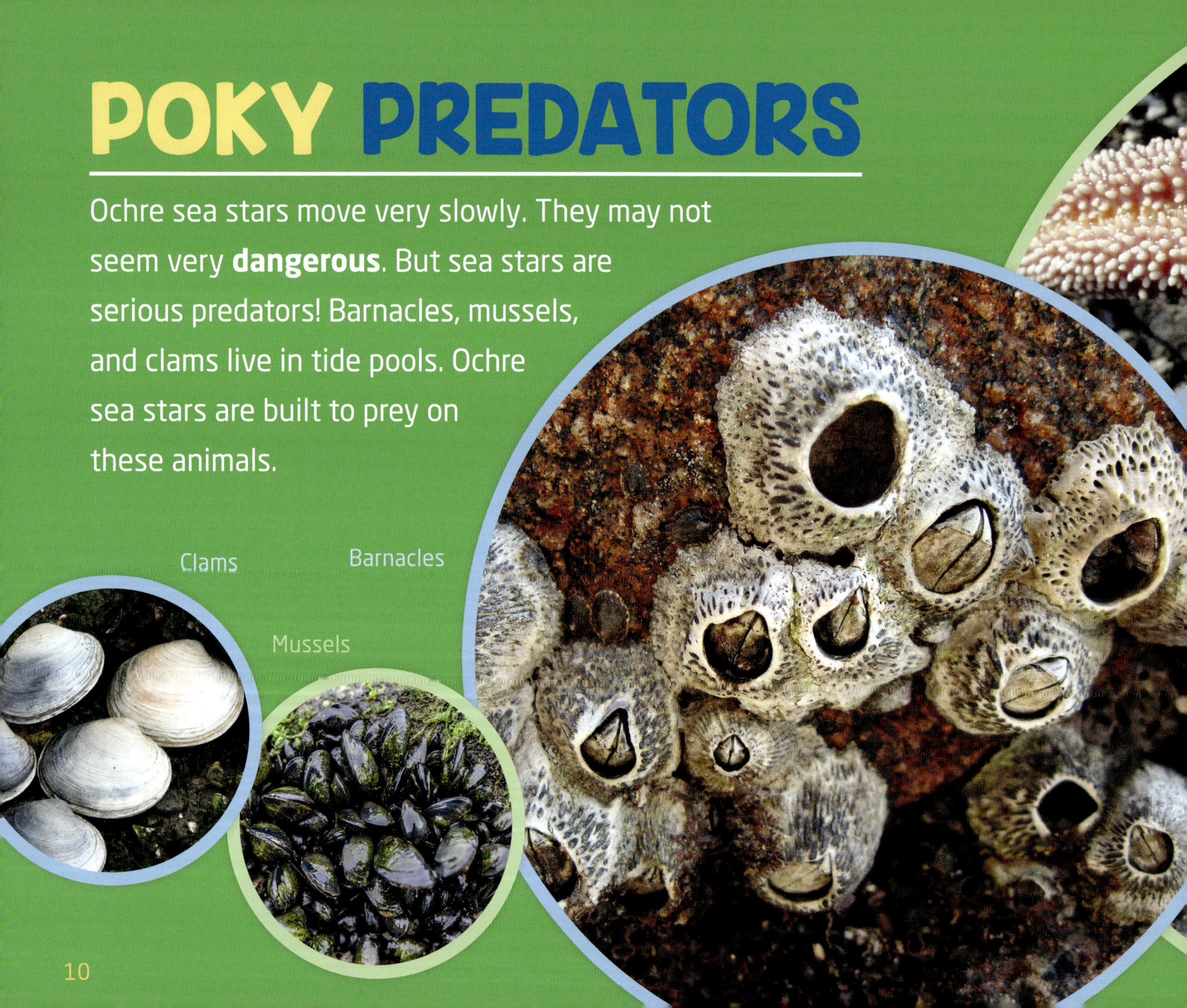

HUNTER'S TOOLBOX

Sea stars have tube feet with tiny **suction cups**. They use these to grip the prey's shell. Then they pull the shells apart.

The ochre sea star can push its stomach outside its body! It sticks its stomach into the opening in the prey's shell. Then it can **digest** the animal inside.

MUSSEL MUNCHERS

Plants and animals compete for space and food in tide pools. Mussels are a big competitor. They cover rocky coastlines and create large mussel beds.

Mussel bed

If there are too many mussels, other animals and plants don't have room to grow. Ochre sea stars are mussels' main predators. They keep mussel populations down.

Ochre sea stars help many different creatures and plants survive in tide pools.

SEA STAR DEFENSE

Ochre sea stars don't have many predators. They have several ways to keep most animals from eating them.

An ochre sea star's skin is covered in tiny, white spines. The spines are tiny **pincers**. The pincers pick off unwanted **algae** and **parasites**.

Sea stars have bad-tasting chemicals in their skin.

Each arm of an ochre sea star has a **sensor** that can sense light. The sensors help the sea star find dark, safe places to hide.

These **defenses** don't work on all predators. Gulls and sea otters prey on ochre sea stars.

Gull

Sea otter

TIDE POOL DINING

Tide pool **habitats** are important for shore wildlife. Tide pools provide food for many different animals. These predators need ochre sea stars to keep tide pool **ecosystems** balanced and healthy.

Sea birds, such as oystercatchers, are the most common tide pool predators.

TIDE POOL PREDATORS

Without ochre sea stars, these animals would have a harder time getting enough to eat.

HUMAN HARM

Ochre sea stars are also affected by human activity that harms tide pool **ecosystems**.

Visitors to the Pacific coast **disturb** tide pools and sometimes remove animals.

Some people collect ochre sea stars. They let the sea stars dry out. Then they sell the sea stars to **tourists**.

Also, pollution from cities and farms runs into the sea. This makes coastal waters dirtier. Dirty water can harm all ocean wildlife.

Sea stars for sale

SICK SEA STARS

Ochre sea stars can get sick. In 2013 and 2014, sea star populations were harmed by sea star wasting **disease**. Scientists believe rising ocean temperatures are one cause of this disease. But scientists still aren't completely sure what makes it so deadly. They are working to find a way to help sea stars.

In just a few days, the disease can turn a sea star into a pile of white goo.

ECO INFLUENCER FACT SHEET

Common name: Ochre sea star

Class: Echinoderm

Life span in the wild: Up to 35 years

Population trend: In recovery

Diet: Carnivore

Size in relation to humans:

FUN FACTS

Sea stars don't have blood. The liquid in their bodies is just sea water.

If they stay in dark, damp spaces, ochre sea stars can live for up to 50 hours out of water.

Sea stars can regrow their arms. So, if an arm gets broken off by a predator, it will grow back!

SEA STAR QUIZ

1. Ochre sea stars are the only kind of sea star.

   True or **false**?

2. What do ochre sea stars eat?

   A. mussels

   B. clams

   C. both A and B

3. What **disease** is harming ochre sea star populations?

Answers: 1. False   2. C   3. Sea star wasting disease

GLOSSARY

algae—plant or plantlike organisms that live in water.

dangerous—able or likely to cause harm or injury.

defense—the act or means of protecting from harm or attack.

digest—to break down food so the body can use it.

disease—a sickness.

disturb—to bother or interrupt.

ecosystem—a group of plants and animals that live together in nature and depend on each other to survive.

habitat—the area or environment where a person or animal usually lives.

influence—to cause something to change.

kelp—a large, brown seaweed.

North America—the continent that includes Canada, the United States, and Mexico as well as other countries.

nutrient—something that helps living things grow. Vitamins, minerals, and proteins are nutrients.

parasite—an organism that lives and feeds on or in a different organism without contributing to the host's survival.

pincer—the claw of an animal such as a crab, or something that is similar to such a claw.

reef—a strip of coral, rock, or sand that is near the surface of the ocean.

sensor—a body part that can sense information such as light, sound, or odor.

suction cup—a flexible cup that sticks when the open end is pressed to a flat surface.

tourist—a person who visits a place for fun or to learn something.